THE

HARNESSING

OF

DRAUGHT ANIMALS

Ian Barwell and Michael Ayre

Practical
ACTION
PUBLISHING

Prepared by Intermediate Technology Transport Ltd.

Ardington

Oxon OX12 8PN

for

Intermediate Technology Industrial Services

Practical Action Publishing Ltd
27a Albert Street, Rugby, CV21 2SG, Warwickshire, UK
www.practicalactionpublishing.org

First published 1982
Printed on demand

ISBN 978 0 946688 72 2

A catalogue record for this book is available from the British Library.

Since 1974, Practical Action Publishing has published and disseminated books and information in support of international development work throughout the world. Practical Action Publishing is a trading name of Practical Action Publishing Ltd (Company Reg. No. 1159018), the wholly owned publishing company of Practical Action. Practical Action Publishing trades only in support of its parent charity objectives and any profits are covenanted back to Practical Action (Charity Reg. No. 247257, Group VAT Registration No. 880 9924 76).

CONTENTS

INTRODUCTION

Role of draught animal power in developing countries

Despite efforts to increase the level of mechanisation, animals remain an important existing or potential source of power in rural areas of developing countries, primarily in agriculture and transport, but also for water pumping, forestry, road construction, etc. The most common method of use is as draught animals, to pull agricultural implements, haul carts, and drive water pumps, though in some parts of the world they are also used as pack animals, carrying loads on their backs.

In certain countries the use of draught animals has declined significantly as a result of increasing mechanisation. However in other places the predicted decline in their use has not occurred. In addition there are countries where efforts are being made to introduce, or to re-introduce, draught animals, either to replace human energy sources in transport and agriculture, or as a consequence of problems associated with the increasing cost and scarcity of petrol and diesel fuel.

Appropriateness of draught animal power

A number of factors affect the appropriateness of draught animals as an energy source in particular circumstances, including:-

- social attitudes towards animals;
- availability of skills in the care and training of animals;

- characteristics and condition of local breeds
  of animal;
- the cost of purchasing animals;
- the availability and cost of fodder;
- the sales value of animals once their working
  life is complete;
- the prevalence of diseases harmful to animals
  (e.g. tsetse fly).

However it is apparent that draught animals will
remain a major source of energy for developing countries
for the foreseeable future, and in some parts of the world
their importance is likely to increase. The current
stock of draught animals in developing countries has been
estimated at about 400 million, the most important species
being bovines (oxen) numbering around 300 million, and
equines (horses, donkeys and mules) about 80 million.

The continuing importance of draught animal power is
reflected in the efforts being made to increase its
effectiveness in developing countries. Many projects are
being undertaken to develop improved animal-drawn
agricultural implements, carts and water pumps, to promote
agricultural systems based on draught animal power, and to
improve the breeding, training, care and management of
draught animals.

Potential for improved harnesses

One aspect of the utilisation of draught animal power
where there is potential for improvement, is harnessing.
The harness used to connect the animal(s) to the drawn
implement is, in effect, the "transmission system" of an
animal-drawn device. Thus the type of harness used has
an influence on the useful power output of the animals and
hence on the working efficiency of particular implements.
This will obviously affect the economic viability of
draught animal technologies.

Simple, wooden yokes for oxen have been used in the developing world for many years. Improvement projects involving the use of draught oxen usually assume the use of these traditional yokes, although there is evidence to show that they are inefficient in transmitting the work output of the animals to the implement, and can in some cases be harmful to the animals, causing a reduction in their useful working life. Innovations have been attempted which indicate that improvements in bovine harnessing are possible.

However relatively little effort has been put into promoting the wide adoption of such improved harnesses, and detailed information about them is not easily available.

Efficient equine harness designs have been utilized for many years in developing countries where draught horses or donkeys are well established. However the design and construction of these harnesses is complex, and relies on the use of leather for the main components. In some countries suitable leather is expensive, or is difficult to obtain. It is also subject to rot and mould attack in very humid conditions. The introduction of horse or donkey-drawn implements to areas where these animals are not traditionally used for draught purposes, and their continued use where traditional harnesses are becoming too expensive, is hampered by lack of information on harness designs utilizing materials and construction methods appropriate to local conditions.

Objectives of the Information Paper

This Information Paper concentrates specifically on draught harnesses for bovines and equines. It is intended to complement the many publications available on other aspects of draught animal power, and to go beyond the catalogue of Indian and African traditional systems usually found in these publications. The Information Paper is directed at people working on draught animal power projects in developing countries, and its objectives are to:-

1. Increase awareness of the role of harnesses in the effective use of animal-drawn implements and the potential for the use of improved harnesses;

2. Provide "state of the art" information on available harness technologies that might be applied in developing countries.

Presentation of Information

The Information Paper is in four parts:

Part 1 - Harnessing Principles

presents information on the draught characteristics of bovines and equines, and discusses the requirements of efficient harnesses.

Part 2 - Bovine Harnesses

describes and analyses traditional bovine yokes and presents information on improved technologies.

Part 3 - <u>Equine Harnesses</u>

describes and analyses existing harnesses and presents
information on designs suited to developing country
conditions.

Part 4 - <u>Information Sources</u>

lists contact organisations and useful publications.

PART ONE: HARNESSING PRINCIPLES

1.1 CHARACTERISTICS OF DRAUGHT ANIMALS

This section summarises information on those characteristics of draught animals which are relevant to harnessing methods. The information is drawn from several publications on the use of draught animals.

Definitions

A variety of animals are used for draught work in different parts of the world but this report concentrates on the two most common types, bovines and equines.

The bovine or ox family includes:

    Bullocks - castrated male oxen
    Cows - female oxen
    Buffaloes - particular breeds of ox
    (Bulls are not usually used for draught purposes).

The equine or horse family includes:

    Horses
    Ponies - small horses
    Donkeys
    Mules - donkey/horse half-breeds

The working performance of draught animals is most usefully defined in terms of draught force and power output. It is convenient to define these at the point where the drawn implement is attached to the harness:

The draught force applied to the implement is
sometimes referred to as the <u>tractive effort</u>.

The power applied to the implement is referred
to as the <u>tractive power</u>.

Working Characteristics

Many factors affect the working performance of draught
animals, including climatic environment, terrain, breed
characteristics, temperament, physical condition, age,
feeding and care, and training and management. It is not
possible to present precise data on the working
performance of animals since this is very dependent on
local circumstances. However the Table below gives an
indication of the level of performance that can be
expected of reasonably well-fed, mature animals.

WORKING PERFORMANCE OF DRAUGHT ANIMALS

Animal	Average weight kg	Approximate draught kgf	Average speed of work m/s	Power developed kW
Bullock	500-900	60-80	0.6-0.85	0.56
Cow	400-600	50-60	0.7	0.35
Buffalo	400-900	50-80	0.8-0.9	0.55
Light				
Horse	400-700	60-80	1.0	0.75
Donkey	200-300	30-40	0.7	0.25
Mule	350-500	50-60	0.9-1.0	0.52

Source: H.J. Hopfen. Farm Implements for Arid and Tropical
Regions FAO 1969.

There appears to be a 'natural average speed' of movement for each breed of animal, which they will tend to adopt wherever possible. This characteristic is reflected in the above figures, which indicate the level of performance that the animal can be expected to maintain over a normal working day. It is possible for animals to generate a higher draught force at a lower speed, or to work at a higher speed but with lower draught force. Neither of these modes of operation will result in an increase in average power output over an extended period.

Draught animals do have the important characteristic that they can generate substantially higher draught forces and power outputs over a short period of time. This allows draught animals, for example, to pull a plough through a difficult piece of ground, to accelerate a loaded cart from rest and haul it uphill, or to uproot a tree. However using draught animals in this way, with intensive short bursts of effort, does reduce their overall working efficiency over a normal working day.

The working performance of a draught animal is a function primarily of its weight, provided that this consists of muscle rather than fat:

Oxen in good condition can reasonably be expected to exert a draught force of about 10% of their body weight.

Horses have a better output:weight ratio than oxen and can reasonably be expected to exert a draught force of about 15% of their body weight.

These relationships are of course dependent on local circumstances. Thus, for example, in parts of Asia where it is difficult to provide adequate fodder for animals, the draught force generated by oxen is typically somewhat less than 10%. In Asia buffaloes are generally larger, heavier and stronger than other bovine breeds.

Using two or more animals harnessed together results in a relative loss in efficiency for each of them, thus:

A pair of animals harnessed together will only produce about 1.9 times the tractive effort of a single animal.

Two pairs of animals harnessed together will only produce about 3.2 times the tractive effort of a single animal.

Three pairs of animals harnessed together will only produce about 3.8 times the tractive effort of a single animal.

In practice it is frequently necessary to use more than one animal in order to generate sufficient total tractive effort to perform particular tasks, even though the tractive effort per animal decreases. The harnessing of two animals together is common. The harnessing of more than two animals to work together is much less common, and makes the harnessing arrangement more complex. As far as possible it is preferable to use a maximum of two animals harnessed together, with the implement matched to their output and to the soil conditions.

Small animals of a particular species tend to have a higher work efficiency, in relation to weight, than large animals. One factor which increases the draught efficiency of small animals is that the "line of pull" is closer to the horizontal than for large animals (this aspect is discussed in more detail on page 9).

The number of hours that an animal can work in a day depends on several factors, including the climate, rest periods and pattern of working. However the Table below, which is synthesised from several different sources gives an indication of the working hours that can be expected.

```
---------------------------------------------
|      WORKING HOURS OF DRAUGHT ANIMALS      |
---------------------------------------------
|  Animal      |  Working Hours/day          |
---------------------------------------------
|  Bullock     |       5-6                    |
|  Cow         |       2-3                    |
|  Buffalo     |       5-6                    |
|  Light Horse |       6-10                   |
|  Donkey      |       3-4                    |
|  Mule        |       8 +                    |
---------------------------------------------
```

Physical Characteristics

Certain physical characteristics of the bovine and equine families are of critical importance in the design of harnesses.

The draught power of equines comes from their strong shoulders and breasts. The strong chest forward of the front legs allows the use of full collar or breast-band harnesses with the draught force applied through the chest. Equines are also strong-backed and are used to carrying substantial loads on their backs. This characteristic allows very effective harnesses to be used for hauling two-wheeled carts.

In contrast to equines, bovines are weak-chested, the draught strength coming from the shoulders. Thus it is generally accepted that, for bovines, the maximum power is exerted if the draught force is appplied from a point just in front of and half-way down the shoulder blades. Furthermore, because bovines are weak-chested the windpipe is exposed. A strap placed around the animal's neck in front of the fore-legs will tend to choke the animal if it is pulled tight. Some breeds of bovine are not traditionally used to carry loads on their backs. This raises questions about the acceptability of using harness arrangements which involve fitting a "saddle" to the animal's back (this issue is discussed in more detail in Part 2). However in Asia buffaloes are commonly used to carry loads on their back.

1.2 HARNESSING PRINCIPLES

There are a series of technical criteria which ideally should be met by a harness design if maximum working efficiency is to be achieved and these are discussed below. However it is self-evident that the successful application of improved harnesses will depend on other factors aparrt from technical efficiency. Some of these factors relate to the broader issue of the appropriateness of draught animal power and are beyond the scope of this Information Paper. Furthermore, the requirements for success in a particular location are best identified by those with local knowledge and experience. However attention is drawn to the following issues:-

Cost

Obviously the cost of a harness should be appropriate to the circumstances of local users. Traditional bovine harnesses are "minimal-cost" designs, and any improvements are likely to involve some increase in cost. The willingness of users to invest in more efficient harnesses will depend on:

- their ability to pay the cost;
- the benefits which they gain from increased working efficiency of their animals. These benefits will depend upon the potential for exploiting this increase in efficiency.

Manufacture

It is obviously desirable that the harnesses should be manufactured using locally available materials and skills. Where an improved harness is being introduced to replace a traditional design, it is desirable that it should be capable of manufacture by the traditional artisans.

Adaption of animals

All animals require training to wear a harness and do draught work. It may be difficult to adapt mature draught animals to working with a different type of harness. In these circumstances it may be preferable to introduce a new harness design with young animals as part of their training programme.

Technical Criteria

1. The harnesses should be durable, and easy to maintain.

2. The harnesses should suit the physical character-
    istics of the breed of animal with which they will
    be used.

3. The harness should not cause discomfort or injury to
    the animal.   This requires:

    - that the harness should be a good fit to the
      particular animal, and smoothly shaped or padded
      so that the loads imposed on the animal's body are
      spread over a large area; and the concentration of
      excessive friction in one place is avoided.

    - that pressure on critical areas of the animal's
      anatomy (e.g. the windpipe) is avoided;

- that the harness should be securely fitted to the animal to prevent it moving about and banging against the animal during normal working and manoeuvering.

4. The harness should be designed to ensure that the animal's strong muscles are effectively utilised, and that it can put its full weight into the work.

5. The harness should be designed to bring the line of draught as close to the horizontal position as possible.   Figure 1.1 shows a bullock drawing a plough using a traditional neck yoke:

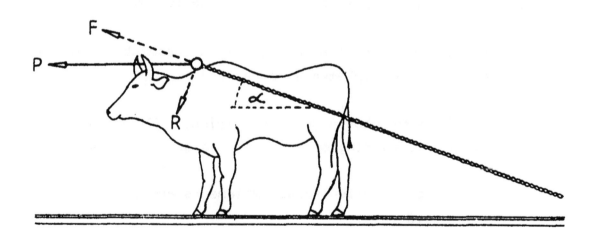

Fig.1.1

The bullock applies a horizontal force P to the yoke.

Because the line of draught from the yoke to the implement is at an angle α to the horizontal the force P applied by the animal is divided into two components:

- the useful draught force F (which is smaller than P) acting along the line of draught.

- a force R pressing the yoke down onto the animal's neck.

Thus only a proportion of the animal's effort is converted into a useful draught force (F), and the animal must also exert some effort to support the force R pressing down on its neck. As the angle of the line of draught to the horizontal (α) decreases, the useful draught force (F) increases, and the force acting on the animal's neck (R) decreases, thereby increasing the draught efficiency.

It is important that the harness should suit the characteristics of the implements to be drawn by the animals. Critical aspects include the following:

6. The harness should have suitable attachment points for the range of implements to be used.

7. The harness should have sufficient 'flexibility' to allow manoeuvering of the implements. For example:

    - turning of agricultural implements are the end of field rows;

    - reversing of wheeled implements, including carts. This requires that the animal should be able to push backwards against the implement.

8. Wheeled implements, often rely on the animal(s) for braking. The harness should be designed to ensure

that the implement does not run forward into the back of the animal(s) and to allow the animals to apply the braking force in a comforatable, efficient way. Ideally such implements should be fitted with brakes but frequently this is not the case. Even when brakes are fitted, the harness should be designed to prevent the implement running forward into the animals in the event of brake failure or handler error.

9. When animals are drawing a two-wheeled cart they act as the third point of support for the cart and therefore carry a proportion of the load. Ideally the cart should be designed, and the load distributed, to minimise the force imposed on the animals. However, even if the cart is perfectly balanced about the wheels when stationary, loads will be imposed on the animals under dynamic conditions (e.g. traversing a bumpy track, accelerating and braking). The harness should be designed so that the downward loads imposed on the animals are supported in an efficient, comfortable way. Ideally the loads should be supported on the animals' backs. In certain circumstances, a two-wheeled cart might tend to tip backwards. The harness should be designed to prevent this, but without causing injury to the animals.

Certain additional considerations have to be taken into account for designs where two animals are harnessed together (double harnesses);

10. While it is desirable for the two animals to be similar in size they are unlikely to be identical. The harness design should allow for some size difference between the two animals without causing discomfort.

11. There should be some flexibility in the harness/

implement system so that, when travelling over uneven ground, there is allowance for vertical movement of the animals relative to the implement. If the system is rigid the animals will experience considerable discomfort.

12. When two animals are harnessed side by side the spacing between them is important. Wherever possible the animals should be close together, but with sufficient space (200-300 mm) between them to allow clearance for the draught chain or rope or draw-pole. However for working between row/ridge planted crops and for the formation of ridges the animal spacing should be twice the nominal row width.

Fig 2.1

PART 2: BOVINE HARNESSES

2.1 TRADITIONAL BOVINE HARNESSES

There are two distinct types of traditional bovine
harness, the <u>neck yoke</u> and the <u>head yoke</u>. The type of
yoke used depends primarily on geographic location and the
physical characteristics of the animals. The neck yoke
is normally used with humped oxen. The head yoke is
better suited to short-necked, non-humped oxen. There is
evidence to suggest that the head yoke is the older form
and that in some areas of the world it was replaced, in
the distant past, by the neck yoke. There is little data
on the relative efficiency of the two types of yoke, but
both suffer from substantive disadvantages. Bovines are
most commonly harnessed in pairs but both types of yoke
can be used for single animals. The head yoke is rarely
used in this way, though it can be useful for light tasks.

Neck Yokes

Since this type of harness is normally used with pairs of
animals the <u>double neck yoke</u> is described first. The
double neck yoke consists of a wooden, or bamboo, beam
placed across the top of the necks of the animals at the
front of the shoulders. The beam is sometimes shaped to
fit the necks of the animals.

A variety of different methods are used to position
the yoke on the animals and to hold it in place:

-   vertical pegs, inserted through the beam on
    either side of each animal's neck, and throat
    straps made of rope, chain or leather passing

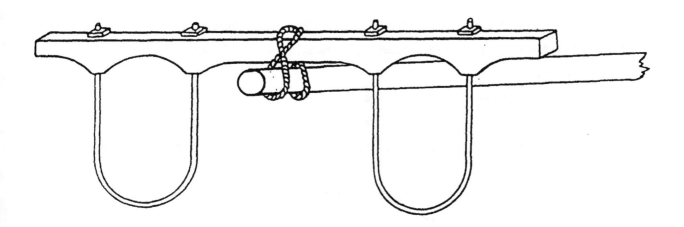

Fig 2.2

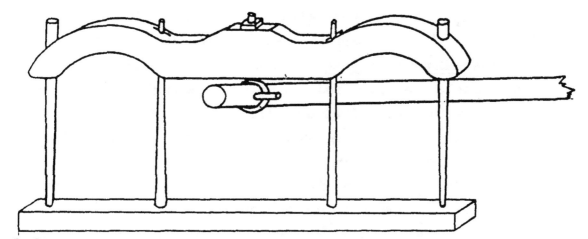

Fig 2.3

- 14 -

under the necks (Fig. 2.1).
- an iron loop for each animal, passing under the neck and attached on either side to the yoke (Fig. 2.2).
- a second wooden beam passing under the animals' necks and connected to the upper beam on either side of each animal by wooden pegs (Fig. 2.3).

The handler controls the animals by means of ropes tied around their head or neck and, often, through the nose.

The dimensions of double neck yokes vary but typically the animals are positioned 1.1-1.2m apart. Implements are attached to the centre of the double yoke beam. Some yokes are fitted with a loop or ring at the centre of the beam to facilitate attachment of the implements. Short beam implements are attached by means of a length of rope or chain which is tied to the yoke (Fig. 2.1) For long beam implements and carts the drawpole is either tied to the centre of the yoke with rope (Fig. 2.2) or passed through a ring and held in place by pegs (Fig. 2.3).

The single neck yoke follows the same principles as the double version and is quite commonly used with buffaloes in parts of Asia. It is quite common for the beam to be made from a shaped piece of wood (Fig. 2.4) which lowers the point of attachment for the implement, gives a better fit on the animal's neck than a straight beam, and improves location and stability.

Short beam implements are connected to a swingletree (also known as a spreader) which is attached to the ends of the yoke by two lengths of rope (Fig. 2.4). Carts drawn by a single ox are fitted with two shafts which are attached to the ends of the yoke (Fig. 2.5).

In some parts of the world, when a single ox is used to draw a cart the strap which normally passes under the animal's neck to hold the yoke in place is replaced by a belly strap positioned immediately behind the animal's

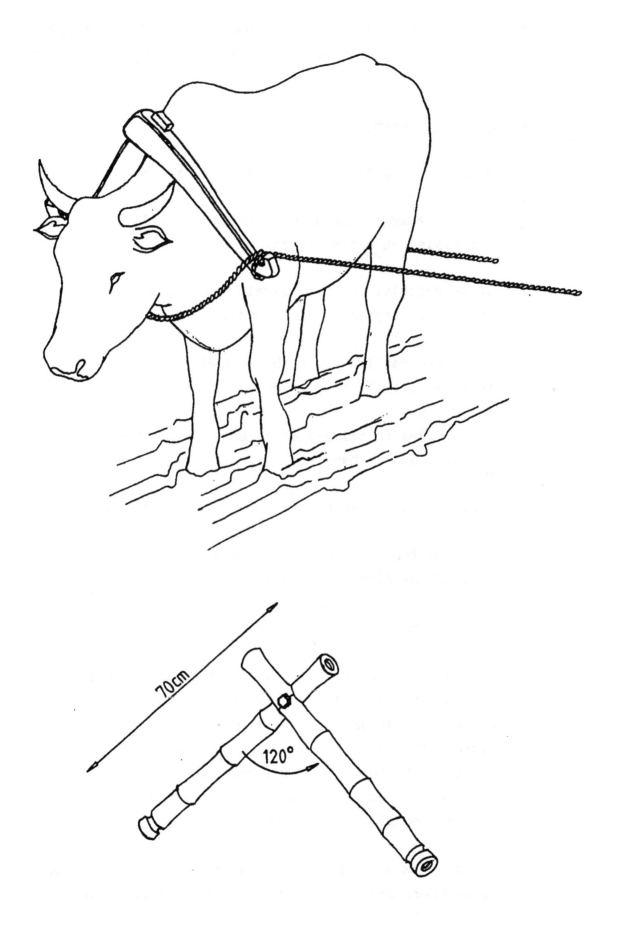

70cm

120°

fore-legs (Fig. 2.5).   This prevents the shafts and yoke
lifting up off the animal but can cause discomfort and
injury.

Assessment of neck yokes

The advantages of the traditional neck yoke are:-

-   it is simple to make, using locally available
    materials, at minimal cost.

-   it is durable, requires little maintenance,
    and can be repaired by the user.

-   users are familiar with the methods required
    to train animals to use the yoke.

-   it provides relatively good control over the
    animals for the handler.

However, the neck yoke has a number of disadvantages:-

-   applying the draught force through the hump
    does not utilize the animals' strength
    efficiently: they cannot harness the full
    strength of the legs, shoulders and back.

-   it can cause injury and discomfort to the animals.
    The load is applied over a small area of the
    animals' necks and, as the yoke is not securely
    fitted, it tends to move about in operation, rubbing
    against the necks. This can lead to wounds and
    parasitic infection which might ultimately reduce
    the working life of the animals.

-   the yoke is pulled backwards when an intensive effort
    is applied, and is restrained by the throat strap.

This is pulled tight against the windpipe.

- the attachment point for implements is high, so
  that the line of draught is at a fairly large angle
  to the horizontal.   As discussed in Part 1 this
  is inefficient, since only a proportion of the
  animals' input is converted into a useful draught
  draught force.

Traditional neck yokes have two specific disadvantages
when used to haul carts:

- since the harness is not securely fitted to the
  animals they cannot apply an effective braking
  force, and it is also difficult to reverse the
  cart.   Under these circumstances the yoke tends
  to slide forwards and hit the animals on the back
  of the head.

- as discussed in Part 1, for a two-wheeled cart the
  animals act as the third point of support.   The
  loads imposed on the animals under dynamic
  conditions can be substantial (often up to 100 kg).
  With the traditional yoke this load is imposed
  on the animals' necks which is not ideal.

There are certain specific features of the double neck
yoke:

- it can compensate for some difference in size
  between the two animals.

- where the drawpole of the implement is attached
  to the yoke through a ring, the yoke has some
  freedom to rotate about the drawpole axis. This
  allows relative vertical movement of the
  animals when traversing an uneven surface, which

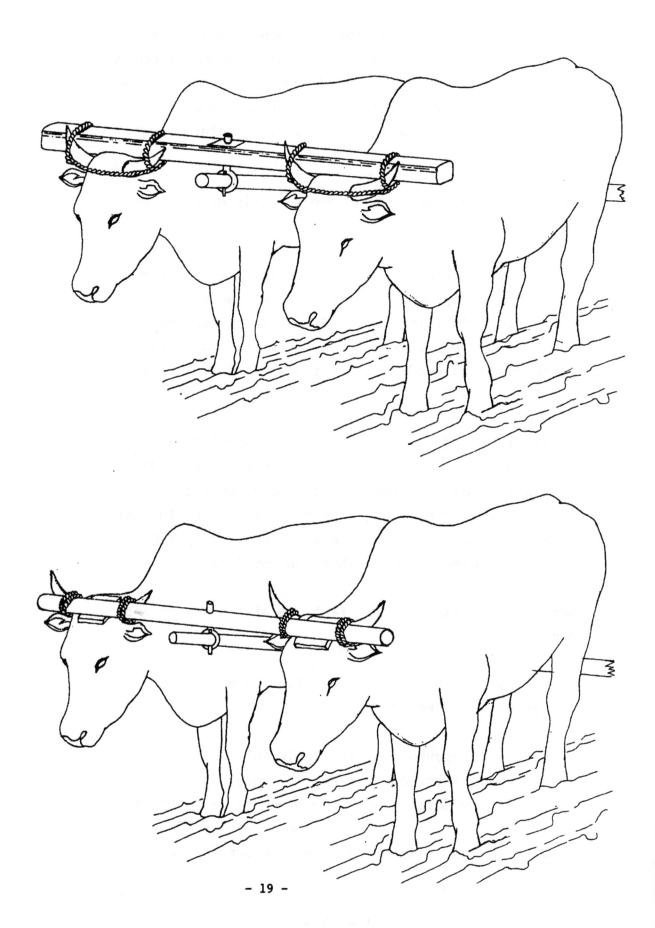

is an advantage.   When the drawpole is tied to
the yoke this freedom of movement is very restricted.

In summary the neck yoke is a minimal cost, simple
device, better suited to humped oxen, which has certain
dis-advantages that limit its transmission efficiency.

Head yokes

Only double head yokes are described since this form of
harness is rarely used with a single animal.

The double head yoke consists of a wooden beam
normally fixed at the back of the animals' heads
immediately behind the horns (Fig. 2.6).   Less commonly
the beam is fitted in front of the horns - a forehead yoke
(Fig. 2.7).   The underside of the beam is hollowed out to
fit over the animals' necks.   The beam is tied firmly to
the animals' heads by means of leather thongs or ropes
passing several times around the horns and across the
forehead.   Notches are sometimes carved on the top side
of the beam to locate the ropes.   Padding is sometimes
provided on the underside of the yoke and on the forehead
to give a more comfortable fit.

Ropes tied to the head or to a nose ring are used by
the handler to control the animals.   Implements are
attached to the centre of the beam using the same methods
as for the double neck yoke.

Assessment of head yokes

The head yoke is normally used with short-necked, non
humped bovines.   The advantages of the head yoke are:-

-   it is simple to make using locally available
    materials.

-   it is relatively cheap.

- it is durable, requires little maintenance, and can be repaired by the user.

- users are familiar with the methods required to train animals to use the yoke.

- because the animals' heads are firmly restrained by the yoke, it provides the handler with good control over the animals, and it is possible to reverse and brake wheeled implements.

However the head yoke has a number of substantive disadvantages:-

- it is not suited to long necked animals. Even for short-necked animals applying the draught force through the neck and the head does not utilize their strength efficiently since they cannot harness the full power from their legs, shoulders and backs.

- it requires more training of the animals than the neck yoke and is also more difficult to fit.

- it can cause injury and discomfort to the animals. The yoke tends to bang against and twist the animals' heads when traversing obstacles or uneven ground though padding will reduce this problem. The yoke can cause breakage of the horns, rendering the animal unserviceable. Unless tied very tightly to the head the yoke will tend to rub, causing sores at the top of the neck.

- the animal can not swing its head (e.g. to shake off flies).

- the attachment point for implements is high, so
  that the line of draught is at a fairly large
  angle to the horizontal. As discussed in Part 1
  this is inefficient, since only a proportion of
  the animals' input is converted into a useful
  draught force.

- because of the rigidity of the yoking system the
  animals experience discomfort, and have to walk
  crabwise, when turning. It also requires the
  animals to be of the same size and to generate
  similar power outputs.

The head yoke has specific disadvantages when used to
haul carts and other wheeled devices:

- while it can be used for reversing and braking
  this is at the expense of discomfort to the
  animals.

- as discussed in Section 1, for a two-wheeled cart
  the animals act as the third point of support.
  The loads imposed on the animals under dynamic
  conditions can be substantial (often up to 100 kg).
  With head yoke this load is imposed at the front
  of the neck. This is not ideal, and in fact is
  less satisfactory than the neck yoke since the neck
  has to support the load as a cantilever.

In summary the head yoke is a low-cost, relatively
simple device, suited to short necked bovines. It has
certain disadvantages that limit its transmission
efficiency.

2.2 IMPROVED BOVINE HARNESSES - GENERAL

In the following sections information is presented on a range of improved harness designs that have been developed or proposed. It is envisaged that the information presented might be used in a number of ways:

- the designs may be adopted directly.

- the design principles may be adopted, but with changes in material specifications to suit local circumstances.

- design concepts may be adopted to develop harnesses suited to specific load circumstances.

Information on harnesses is presented in two parts:

- harnesses for single animals

- methods of harnessing pairs of animals to an implement.

Although the use of single bovines for draught work is less common than the use of pairs of animals, information on single harnesses is presented first for the following reasons:

- with the introduction of improved harnesses and implements there is considerable potential for using a single animal where a pair are required at present.

- the design principles behind many of the improved harnesses are also applied to double harnesses.

-   a pair of animals can be <u>independently</u> hitched
    using single harnesses, to an implement.

This section discusses the general issues that relate
to the design and introduction of improved harnesses. It
is clear from projects undertaken in different parts of
the world, and for which detailed test data is available,
that improved harnesses can give a significant increase in
the work output of bovines.   It is difficult to make
rigorous comparisons between test results from different
projects because of variations in conditions and the types
of trials carried out.   Tests on improved harnesses fall
into two broad categories:

-   short term, controlled tests to establish
    maximum power output, maximum draught force
    etc.

-   longer term tests under 'normal' working
    conditions to establish useful energy output,
    useful work done, etc.

Improvements in performance of up to 60% have been
obtained from such tests.   The other major objective
behind the development of improved harnesses, apart from
increases in work performance, has been to reduce injuries
which can shorten the working life of animals.   Because
the benefits of such measures are essentially long term no
testt data is available on this aspect.   Inherent in much
of the work on improved harnesses is the objective of
providing designs which, in contrast to traditional yokes,
are well fitted to the animals, and spread the loads
applied to the animals over as great an area as possible.
Thus many of the designs presented include padding and/or
provision for adjustment.   This approach has certain
important implications:

1. The positioning of the elements of the harness is critical since free movement of the limbs, and breathing and swallowing must not be restricted.

2. Where available, dimensional information on improved harness designs is presented. However this should be used with caution since in practice dimensional details must be defined to suit the characteristics of local breeds of animal. The provision of a degree of adjustment of the harness, or of soft padding, can allow a single size of harness to fit a limited range of sizes of animal. However, a 'universal' harness which fits a wide range of animals is unlikely to be practical, unless the harness is very simple.

3. Evidence on the ease with which bovines will adapt to a new type of harness is conflicting. On the one hand there are examples of animals used to working with a neck yoke adapting immediately to the very different, and more complex, three-pad collar harness. On the other hand, some attempts to develop improved harnesses have failed because mature animals were unwilling to wear them. Clearly, many factors will affect the acceptability of a new harness, including the degree of difference from traditional practice, and the age and temperament of particular animals.

4. Several of the improved harness designs incorporate padding, and specific materials are recommended for this purpose. There is scope for using alternative locally available materials for the padding itself and for the covering, but the latter must be comfortable for the animal. Similarly, where straps or belts are incorporated into the harness a range of material options is possible. Materials for straps are discussed in more detail in Part 3: Equine Harnesses.

5. Where a harness design incorporates padding, then there must be sufficient quantity of padding to ensure that, when draught loads are applied to the harness, the hard wood and metal parts do not press against the animal.

6. The use of harnesses with additional straps and provision for adjustment means that the fitting of the harness to the animal(s), and its removal, will be more complicated, and take longer than with the simple, traditional neck yoke.

Note: In the illustrations of improved harnesses in the remainder of this part of the report, a standard drawing of a 'generalised' humped bovine has been used, in order to clarify the descriptions of the harnesses. However some of the harnesses can be used with non-humped bovines, and where this is the case it is noted in the text.

2.3 SINGLE BOVINE HARNESSES

The following pages of this section present information on a range of single bovine harnesses. The harness designs are described first, and then the different methods of attachment to short beam agricultural implements and wheeled implements, including two-wheeled carts, are discussed.

Four basic categories of harness are described:

adaptions of the neck yoke

- Chinese V-yoke
- Swiss withers yoke

These are relatively simple harnesses based on the

principles of the traditional neck yoke but offer improved comfort location and utilisation of the animal's strength. They can be designed to suit the physical characteristics of all working breeds of bovine.

flexible harnesses

- European flexible harness
- Zimbabwe flexible harness

These are relatively simple harnesses which have no rigid parts. Compared with the traditional neck yoke they offer improved comfort and utilisation of the animal's strength, and are best suited to humped bovines. They are not suitable for drawing wheeled implements.

collar harnesses

- Three-pad collar harness
- Allahabad collar harness
- Vietnamese collar harness

These are relatively complex, expensive harnesses but appear to offer the greatest increase in efficiency compared with the traditional neck yoke. They can be designed to suit the physical characteristics of all working breeds of bovine.

The above three categories of harness all represent attempts to bring the point at which the draught force is applied closer to what is generally accepted as the ideal position, i.e. just in front of, and halfway down, the shoulder blades.

back harness

- Japanese back harness

This harness is based on the quite different principle of applying the draught force through the animal's back. It is relatively complex, but offers improved comfort and efficiency. It is only suited to those breeds of bovine which will accept a load being placed on the back.

SINGLE BOVINE HARNESSES

Chinese V-Yoke
Swiss Withers Yoke
Flexible Harness
Zimbabwe Flexible Harness
Three Pad Collar Harness
Vietnamese Collar Harness
Allahabad Collar Harness
Japanese Back Harness

DOUBLE BOVINE HARNESSES

IDC Clarkson Ox Yoke
Double V-Yoke
Savar Padded Yoke
Mymensingh Yoke
Allahabad Yoke

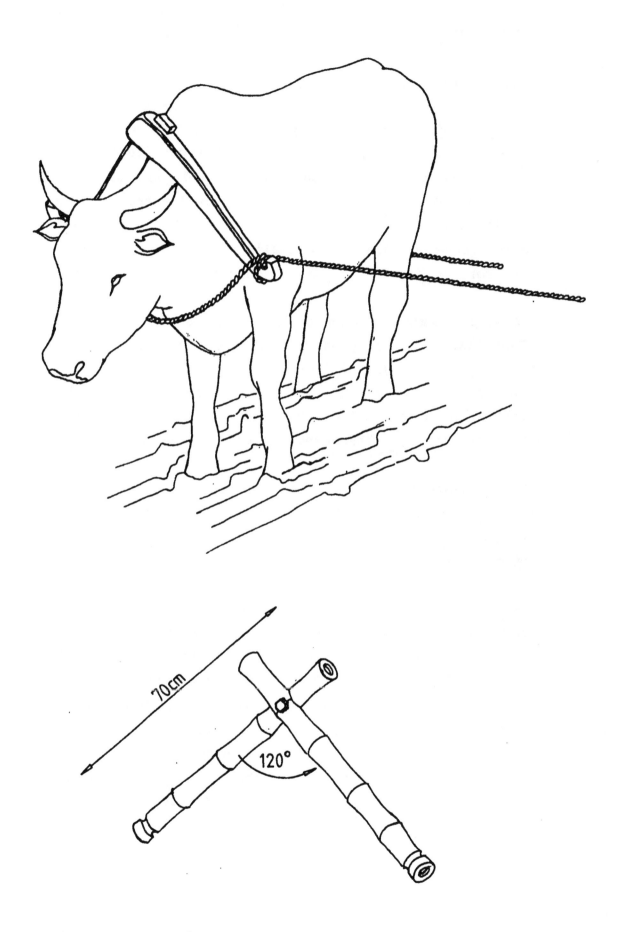

CHINESE V-YOKE

Source: Used in China

Description of Design: The yoke consists of two pieces of wood, joined at an included angle of about 120°.   The underside of the yoke should be carved and smoothly rounded to provide a good fit on the animal's neck.   The yoke fits over the animal's neck.   The yoke is retained by a rope passing under the animal's neck.   The traces or shafts for the implement are attached to each end of the yoke.

A simple bamboo version of the yoke has been developed in Tanzania.   Two lengths of bamboo are bolted together with an included angle of 110-125°.   It was found that if the angle was any smaller the yoke tended to rub on the animal's shoulders.   Grooves formed at each end of the bamboo yoke provide attachment points for traces.

Status of design: The V-yoke has been widely used in China for many years.   No test data is available on its efficiency in comparison with traditional neck yokes used in other developing countries.

Assessment of Design: The V-yoke is simple and cheap to produce, and can easily be made to be a good fit on a particular animal.   It can be used with any breed of bovine.   Compared with the traditional neck yoke it offers the advantages of a lower hitching point, better utilisation of the animal's strength, and, because it fits on to the sides of the neck, improved location and stability.   The yoke might be improved by the use of padding where it is in contact with the shoulder.

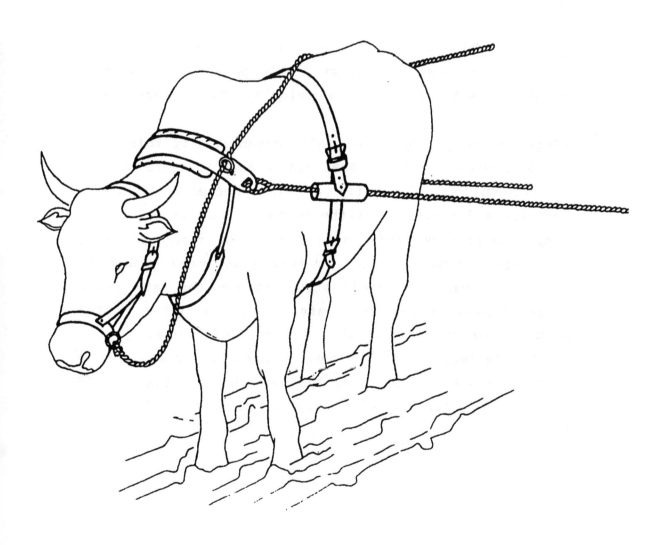

SWISS WITHERS YOKE

Source: Formerly used in Switzerland and other parts of Europe.

Description of Design: The yoke is made from a strip of flat steel plate bent roughly to the shape of the animal's neck, and padded. It is held in place by a loose fitting metal bar, a strap or combination of the two, passing under the animal's neck. Attachment points are provided at each end of the steel plate for the traces. Straps around the back and belly, and sometimes across the hips of the animal, locate the traces.

Status of design: The Withers yoke was widely used in Europe. No test data is available on its efficiency in comparison with traditional developing country neck yokes.

Assessment of Design: The major advantages of this yoke, compared with the traditional single neck yoke are that it is padded, and the belly and hip straps give better location and stability of the harness. It is relatively simple to produce, and can be made to be a good fit on a particular animal. Wood, rather than steel, could be used to provide the rigid structure for the yoke and a range of padding and strap materials are feasible. The Withers yoke can be used with any breed of bovine.

The features of the Chinese V-yoke, combined with the padding, and trace guides of the Swiss Withers yoke, would appear to offer a relatively simple, but effective harness.

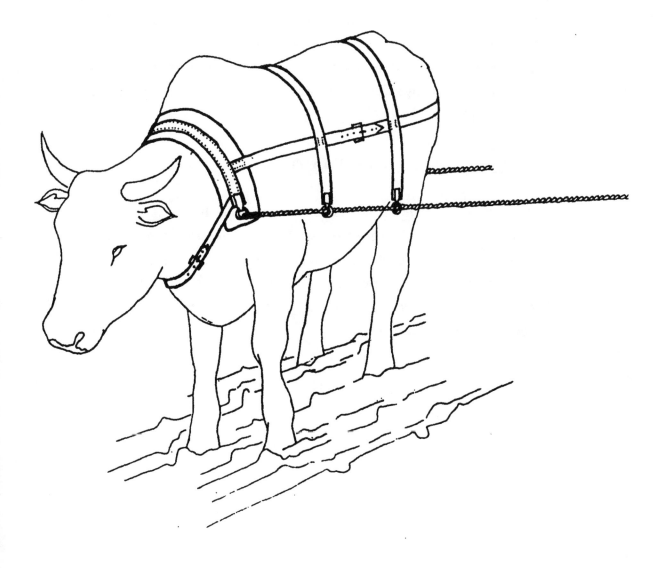

FLEXIBLE HARNESS

<u>Source</u>: European

<u>Description of design</u>: This harness, which has no rigid parts, is made up of a series of leather straps. The main, broad strap, often reinforced with a narrower strap, fits across the neck and above the shoulders with metal rings at each end to which the traces are attached. This shoulder strap is held in place by an adjustable strap passing under the animal's neck. Additional back and hip straps locate the traces and are attached to longitudinal straps which end in a loop fitting around the animal's tail. These straps prevent the harness slipping forward on the animal.

<u>Status of design</u>: was used in Europe.

<u>Assessment of design</u>: Manufacture of the harness requires the availability of leather and leather-working skills. The harness is comfortable, well located, easily adjustable to fit a particular animal, and provides a relatively low hitching point. The flexible harness is best suited to humped oxen.

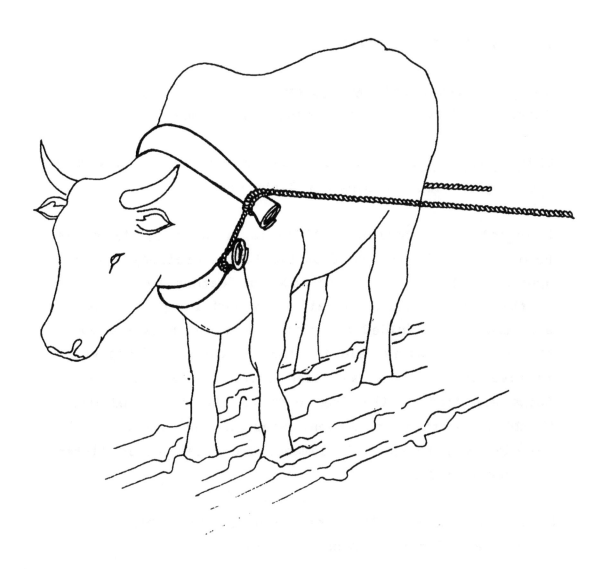

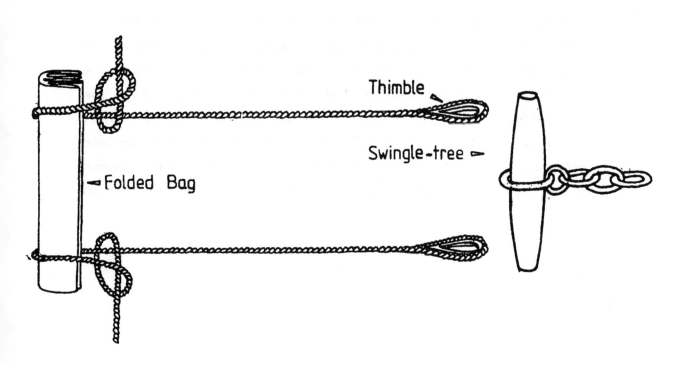

Thimble

Swingle-tree

Folded Bag

ZIMBABWE FLEXIBLE HARNESS

Source: Developed by Appropriate Technology Section, Institute of Agricultural Engineering, Zimbabwe.

Objectives of design: to provide an effective, inexpensive single animal harness for light work.

Description of design: This is a simple, flexible harness.   It is made up from two 100 kg grainbags and two approximately 3m lengths of rope or leather thong.   One of the bags is folded in half twice and a rope is tied to each end.   The folded bag fits across the neck and above the shoulders and is held in place by the second bag which is also folded twice and tied under the animal's neck. Loops are tied at the free ends of the ropes and metal thimbles inserted to prevent wear of the rope.   The swingle-tree, to which implements are attached, is fitted into the thimbles.

Status of design: No information is available on the performance of this harness.

Assessment of design: The harness is cheap, uses locally available materials, and could be made by the farmer.   It should be more comfortable for the animal than the conventional wooden yoke and provides a relatively low hitching point.   It is best suited to humped oxen. However, in the absence of belly and backstraps it is not as well located on the animal as the flexible harness described on the previous page, and may tend to twist in use.

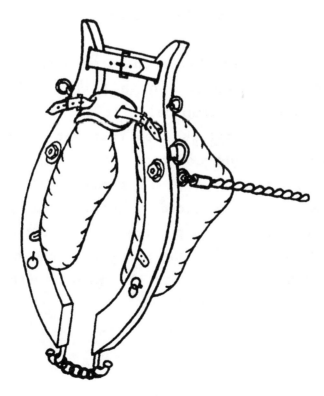

THREE-PAD COLLAR HARNESS

Source: used in Europe

Description of Design: The harness is made up of two
shaped wooden hames (the rigid frames of the collar)
connected at the top by an adjustable leather strap and at
the bottom by a chain.   Large, shaped, leather-covered
pads are attached to the hames where they fit in front of
the shoulders. A third, smaller pad, fitted between the
two hames and adjustable by means of two leather straps,
sits on the neck of the animal in front of the withers.
Attachment points for the traces are fitted at
approximately the mid-points of the hames. An adjustable
strap, around the back and belly of the animal, locates
the traces.

Status of design: Introduced in the late 1930's it was
more effective than, and soon replaced, other types of
cattle harness in most central European countries.
Testing of a three-pad collar harness for logging work has
recently commenced in the Philippines.   Initial results
show a significant improvement in tractive efficiency
compared with the traditional yoke.

Assessment of Design: Based on horse-harnessing practice,
it provides the advantages of a full collar harness but
does not impede the windpipe.   It is adjustable to fit a
particular animal, allows free movement of the shoulders
and a natural straight position of the back by which means
the animal is able to apply its power effectively.   It
can be used with any breed of bovine but is relatively
complex and expensive to produce.   In particular careful
attention is required to the shaping of the pads to ensure
they fit comfortably against the animal without impeding
movement of the shoulders.

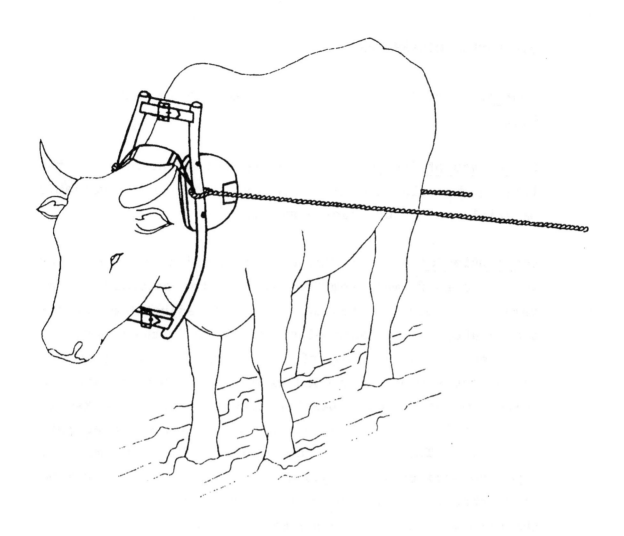

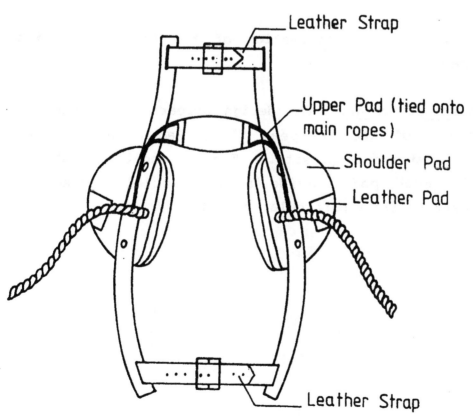

Leather Strap

Upper Pad (tied onto main ropes)

Shoulder Pad

Leather Pad

Leather Strap

VIETNAMESE COLLAR HARNESS

Source: Developed by National College of Agriculture, Vietnam.

Objectives of design: to exploit the principles of the three pad collar harness, but utilize locally available materials and manufacturing skills.

Description of design: The harness consists of two curved wooden hames formed from 20-40mm diameter hardwood. The hames are joined at top and bottom by adjustable leather straps attached by means of slots formed in the hames. A shoulder pad, consisting of a 15-20mm thick plywood base and a canvas covering padded with foam rubber or kapok is bolted to each hame. Leather rubbing pads are tacked to the outer face of each shoulder pad where the traces pass over it. This prevents the traces, which are made of rope and spliced to the hames, from wearing and breaking where they pass over the pads. The neck pad is tied to the traces where they attach to the hames.

Status of design: Utilisation of the harness and its performance are not known.

Assessment of design: The design appears to retain the advantages of the European Three Pad Harness while reducing cost and complexity. The top and bottom straps and the neck pad provide a substantial degree of adjustment to fit a particular animal.

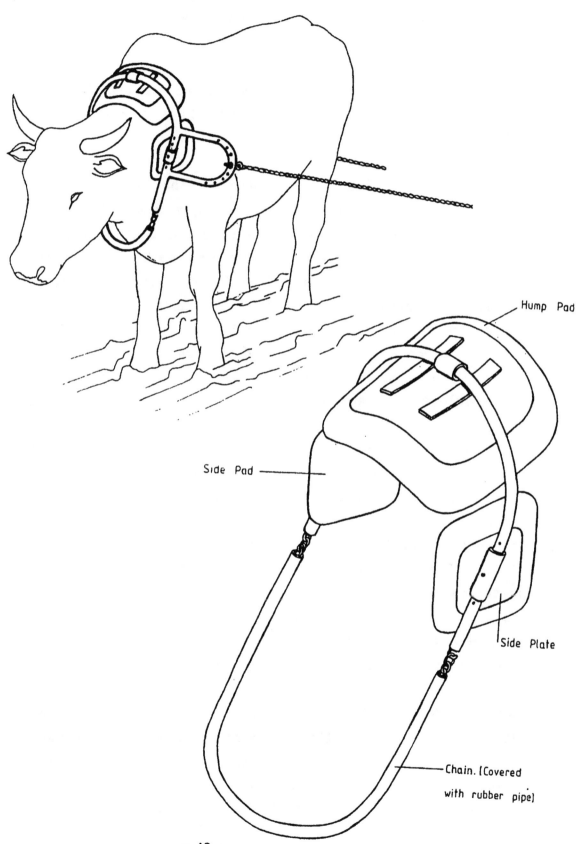

Hump Pad

Side Pad

Side Plate

Chain. (Covered with rubber pipe)

- 42 -

ALLAHABAD COLLAR HARNESS

Source: Developed by the Agricultural Implements and Power Development Centre, Allahabad Agricultural Institute, India 1964.

Objectives of design: to improve power utilization and reduce prospects of injury compared with traditional yokes.

Description of design: The basis of the harness is an inverted U-piece made from 25mm o.d. x 3mm thick steel tube.   The straight ends of the tube are bent forward at 30°.   Shoulder pads are attached to each side of the U-tube.   The pads are made from 2.5mm thick steel plate covered with canvas and padded with coconut coir.   Tubes welded to the back of the pads fit over the straight ends of the U-piece and are located in a range of positions by pins.   The neck pad is constructed in the same manner as the shoulder pads and attached to the top of the U-piece in a similar manner.   The neck pad is not adjustable but has limited freedom to pivot.   Excessive movement is prevented by two steel strips welded to the underside of the U-piece.   The ends of the U-piece are connected by a length of chain, covered with rubber pipe, passing under the animal's neck.   Additional U-tubes welded to each arm of the main U-piece, provide a range of attachment positions for the traces, and ensure that they do not rub against the animal's sides.   However, for heavy-duty haulage of a two-wheeled cart, the cart shafts are attached directly to the main U-piece.   (See page 49)

Status of design: Detailed, controlled tests were carried out to compare the collar harness with a Japanese back harness (see later pages) which had been found to be more efficient than a traditional neck yoke.

Under continuous working conditions, the collar harness gave about 14% more power than the Japanese harness and allowed the animal to work about 12% longer without any significant drop in power output. Under a normal working schedule, the collar harness gave about 6.5% more power and 16.5% higher draught than the Japanese harness. Compared with the Japanese harness, use of the collar harness reduced the time required to plough a given area of land by about 20%. As far as is known, no concerted effort has been made to disseminate the design.

Assessment of design: The design is based on the same principles as the European Three Pad Harness, and provides some degree of adjustment to fit a particular animal. The harness is relatively simple to make provided that basic metal-working facilities are available. At the time of development the harness was estimated to be about three times the cost of a traditional neck yoke. The test results show the design to be significantly more effective than the Japanese back harness, which is itself more effective than the traditional yoke. Like other collar harnesses it can be used with any breed of bovine.

JAPANESE BACK HARNESS

Source: used in Japan.

Description of design: The harness consists of a wooden saddle frame which fits over the animal's back and is padded by means of a saddle cloth. The saddle is V-shaped so that it locates on the sides of the animal's back and does not press on the backbone. The saddle is strapped on by means of a belly band and is prevented from slipping forwards or backwards by a loosely fitted throat strap and breech strap. The traces are joined to form a single 'loop' of rope which passes across the saddle and is retained by fixing poles at the ends of the belly band. This has the effect of pulling the saddle down onto the animal's back as it draws the implement, to prevent the throat strap from tightening.

Status of design: This type of harness was widely used in Japan. No quantitative test data on its performance is available but trials in India suggest that it is more effective than the traditional neck yoke.

Assessment of design: This type of harness is quite different from other designs. It is based on the principle that the point of traction should be in vertical line with the centre of gravity of the animal. Since it is of predominantly wooden construction it should be relatively easy and cheap to manufacture locally and, as it is padded, is comfortable for the animal. It is adjustable to suit different sizes of animal. It could be used with both humped and non-humped bovines though some breeds may not accept the fitting of a back saddle. It has the specific advantage when hauling a two-wheeled cart that the cart load is applied to the animal's back, rather than its neck.

<u>Note</u>:

Another version of this harness uses a simple neck yoke instead of the breast strap.   The ends of the yoke are tied to the saddle.   The yoke is used primarily to prevent the saddle slipping backwards and the draught force is applied predominantly through the saddle.

Single Harnesses - Attachment of Implements

1. Wheeled Implements

The connection between the wheeled implement and the harness must be rigid, to prevent the wheeled implement running forward into the back of the animal. Flexible harnesses are not considered to be suitable for use with wheeled implements.

For two-wheeled carts which are likely to be the most common wheeled implement drawn by a single bovine, and which often carry heavy loads, an additional consideration is that the animal is the third point of support for the loaded cart.

The simple method of attaching a two-wheeled implement to a single harness is by means of two shafts. These are rigidly fitted to the implement (and, in the case of a cart may form part of its structure) and run forward on either side of the animal, to the trace attachment points. The shafts should not rub against the sides of the animal, and it is therefore often desirable for them to be angled inwards towards the front when viewed from above. The shaft length should be set to give sufficient clearance between the animal and the implement for its rear legs to move freely.

The above method of attachment is suitable for all the neck yokes and collar harnesses described, and for the Japanese back harness. In the case of the Japanese back harness, since the attachment points are on the saddle frame, the loads imposed on the animal by a two-wheeled implement are applied across the back. In the case of the neck and collar harnesses the loads are applied to the neck. As discussed in the section on traditional neck yokes this is not entirely satisfactory for two reasons:

(i)   the harness tends to slide forward when reversing

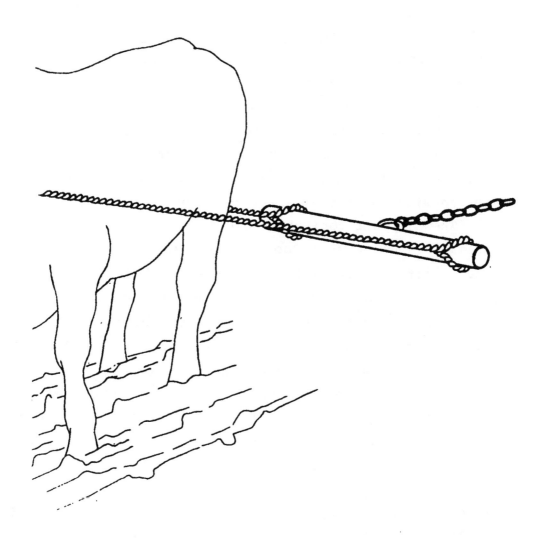

or braking (though this can be overcome by linking the harness to a hip or "breech" strap, as illustrated for the European Flexible Harness.)

(ii)   the neck is not the best place to support the vertical load imposed by the wheeled implement or cart.

   For those breeds of bovine that find it acceptable it is preferable for the load to be applied to the back. This can be achieved by the addition of a saddle to support the shafts, with the draught force being applied through the neck or collar harness, as detailed below.
   However, if this method is used, the cart must be well balanced, since excessive downward pressure on the spinal column will cause the animal to 'sink'.
   The simplest form of saddle is a broad belt across the animal's back with a loop at each end through which the shafts fit.   A belly band is added to hold the saddle in position.   The shafts are attached to the trace mounting points by means of lengths of rope or chain.   The shafts are free to slide backwards and forwards in the loops but should be of such a length that they cannot drop out of the loops.   The addition of a breech strap attached either directly or via the saddle to the shafts assists in braking and reversing the wheeled implement and prevents it running forward into the animal.
   The load imposed on the animal's back can be spread over a wider area by supporting the back belt on a wooden saddle frame sitting on a padded saddle-cloth (as used on the Japanese back harness).
   A further modification is to connect the neck or collar harness directly to the implement by means of traces, so that the shafts do not transmit the draught force but simply transfer the implement load to the animal's back.

2. Short-Beam Agricultural Implements

It is difficult to pull long-beam agricultural implements satisfactorily with a single animal harness.   The normal practice is to use short-beam implements attached to the harness by means of a "swingle-tree" or "spreader" bar, and traces.   The traces are usually made of rope or chain,* though leather thongs can be used.   The traces attach to the draught harness and pass back on either side of the animal to the ends of the swingle-tree.   Some of the single harnesses described have back and/or hip-straps to guide the traces.   The swingle-tree should be sufficiently wide to prevent the traces rubbing on the animal's sides.   The traces should be sufficiently long to give some clearance between the rear of the animal and the swingle-tree to allow:-

(i)   free movement of the animal's rear legs

(ii)   some "bouncing" of the swingle-tree when the
       draught force is applied or removed

The implement is attached to the centre of the swingle-tree by means of a rope or chain.

The swingle-tree is normally made of wood.   The traces are tied directly to the swingle-tree or attached to rings or hooks fitted to the swingle-tree.   This method of implement attachment is suitable for all the single harnesses described.

* If chain traces are used they should be of the "closed-
  link" type.   Open-link chains tend to have sharp edges
  which can harm the animal.

2.4 DOUBLE BOVINE HARNESSES

This section presents information on harnesses that can be used with pairs of animals. There are two approaches to harnessing pairs of animals:

1. Double harnesses following traditional practice where the two animals work side-by-side in a rigid harness. Several such designs are described, together with methods of attaching the implement to the harness. The designs described fall into three categories:

   adaptions to the traditional double neck yoke

   - IDC-Clarkson ox yoke
   - Double V-yoke

   padded adaptions of the double neck yoke

   - Savar padded yoke
   - Mymensingh yoke

   rigidly linked collar harnesses

   - Allahabad yoke

   this appears to offer the greatest increase in efficiency compared with the traditional neck or head yoke, but is also the most complex and expensive.

2. Independent hitching of two animals, each fitted with a single harness, to the implement. The range of single harnesses is described in the previous section, and the methods of attachment to the implement are covered here. Using the same principles it is also possible to hitch three or more animals independently

to a single implement.

The common practice at present with bovines is to use a double harness, though in parts of China independent hitching is used. However independent hitching is the predominant method used when two or more horses are harnessed to a single implement. One advantage of independent hitching is that it is a simple task to alter the spacing between the two animals for different draught applications. This spacing may not be appropriate for other operations such as ploughing or hauling a cart. With double harnesses the spacing of the animals is controlled by the width of the harness. Thus a separate harness would be needed for each application requiring a different animal spacing. However a double harness does simplify control of the animals since they are constrained by the rigid connection between them. With independent hitching the animals have much greater freedom of movement and the control system must ensure that they work in unison.

IDC Clarkson Ox Yoke

Source: Developed by the Industrial Development Centre, Samaru, Nigeria, 1969.

Objective of design: to reduce discomfort to animals, and to improve transmission efficiency of their pulling power.

Description of design: The yoke consists of a shaped smoothly finished piece of timber about 125mm wide. The areas of the yoke which bear on the animals are contoured to the shape of the necks. The centre section of the yoke is below the neck sections to lower the line of draught for improved efficiency. The draw ring can slide along a loop attached to the underside of the yoke. As the animals start to pull the implement, the draw ring slides to the back of the loop, tilting the yoke to fit into the slope of the neck. The yoke is retained in position by two loops made of iron pipe which are angled forward and adjustable for position by means of spring pins to clear the animals' throats.

Status of design: no test data is available on the efficiency of the yoke or on the level of adoption of the design.

Assessment of design: in the absence of any test data the design appears to offer some improvement over the traditional yoke in terms of both comfort and pulling efficiency.

The yoke is relatively simple to manufacture and uses easily available materials and skills. It would be slightly more expensive than the traditional yoke. The yoke is best suited to humped bovines.

Double V-yoke

No information has been obtained on this type of yoke but it would seem feasible to form a double yoke by rigidly connecting two of the single V-yokes described in the previous section.   Compared with the traditional double yoke this would offer the advantages of better location of the yoke on the animals and a lower hitching point.   Such a yoke would be relatively simple and cheap to manufacture locally.   It is also feasible to add padding to a Double V-yoke.

Savar Padded Yoke

Source: Developed by the UNICEF-supported Savar Village Technology Action Programme in Bangladesh, 1978.

Objectives of design: to reduce the problem of neck sores caused by the traditional yoke.

Description of design: a simple, low-cost adaption of the traditional bamboo beam yoke used in Bangladesh. The yoke is padded where it sits on the animals necks in order to spread the load and minimise the sores caused by rubbing of the yoke. The adaption consists of two 'tubes' of burlap cloth stuffed with raw jute, fitted over the standard yoke and retained by strips of tin tacked to the yoke at each end of the tubes. No other modifications are made to the traditional yoke design.

Status of design: There is only limited data on the effectiveness of the improved design. Some improvement in the condition of neck sores on animals was reported after the padded yoke was introduced, but no data on its long-term effect is available.

Assessment of design: The adaption is aimed at overcoming one specific deficiency of the traditional design, its tendency to cause neck sores. There is some initial evidence of its effectiveness in doing this. The adaption would increase the cost of the traditional yoke, but the increase in cost would not be large. The materials used are cheap, easily available, and the adaption could be made by the farmer himself or by a village carpenter or blacksmith. Alternative "padding" materials could be used in areas where jute is not available. One advantage of the adaption is that it can be applied to existing yokes.

Note: The Dunlop company of India, which manufactures pneumatic tyred bullock carts, offers a yoke design based on the same principle, namely a straight beam (in this case made of tubular steel) padded where it sits on the animals' necks.

Mymensingh Yoke

Source: Developed by the Department of Farm Power and Machinery, Bangladesh Agricultural University, Mymensingh, Bangladesh, 1979.

Objectives of design: to allow the animals to push with the shoulders as well as the neck, and provide a larger, padded bearing area for the animals.

Description of design: The experimenters evaluated three designs, all based on the collar principle. All three designs are illustrated and the most successful is described in detail. It consists of a wooden beam with inverted U-shaped wooden frames bolted at each end. These frames are padded with cotton and covered with leather. Each frame is retained in position on the animal by means of a rope passing under the neck.

Status of design: A series of tests were carried out to compare the improved yoke with two traditional Bangladeshi types. Short term controlled tests under different conditions showed that the improved yoke gave an average increase of 17.5% in maximum draft output and 11% in maximum power output compared with the better of the traditional yokes. Under longer term testing the improved harness gave an 18% increase in useful energy output over a 5-hour period.

Assessment of design: The Mymensingh yoke appears to offer a significant improvement in efficiency over the traditional neck yoke. The padding should also reduce the occurrence of neck sores. The yoke is suitable for local manufacture using available materials and skills. Different padding materials could be used, according to availability. The improved harness is estimated to cost about 100 Taka (U.S.$6) more than traditional types.

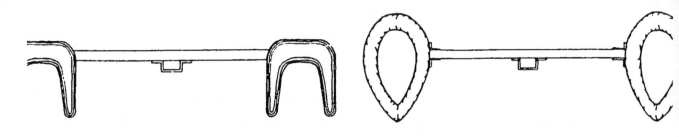

Type 1:   Wooden 'collar'   Type 2 : Padded full collar

Type 3 : Preferred design

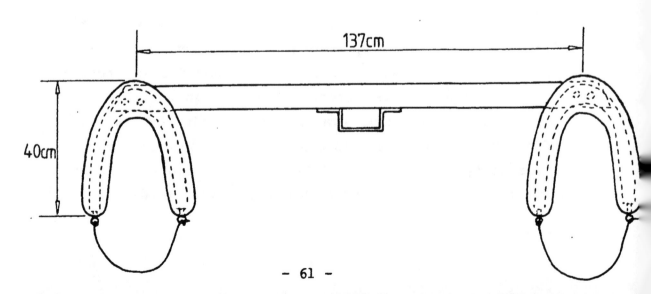

137cm

40cm

Allahabad Yoke

Source: Developed by the Agricultural Implements and Power Development Centre, Allahabad Agricultural Institute, India, 1964.

Objectives of design: to develop a more efficient harness to improve the utilisation of animal power and also reduce prospects of injury.

Description of design: The design is based on the Allahabad single collar harness described in the previous section. It consists of two of the single collar harnesses rigidly connected together by means of two horizontal steel tubes of 31mm o.d., the upper one having a wall thickness of 3.8mm, the lower one 3.2mm. These tubes are welded to the main U-pieces of each collar. The two tubes are braced by two lengths of angle iron spaced 120mm apart. Angle iron brackets bolted to these braces provide a hitch point which can be adjusted to different vertical positions. The hitch point is lower than on traditional yokes.

Status of design: A comprehensive series of tests were carried out to compare the performance of the harness with one of the more efficient traditional designs. Under continuous working conditions, the improved harness gave about 14% more power than the traditional yoke and allowed the animals to work 30% longer without any significant drop in power output. Under a normal working schedule, the harness gave 38.5% more power than the traditional yoke. Use of the improved harness reduced the time required to plough a given area by 23%. There is no evidence of widespread adoption of the design.

Assessment of design: The Allahabad yoke clearly gives a

substantial increase in efficiency compared with
traditional designs and minimises the risk of neck sores.
The yoke is relatively simple to make provided that basic
metalworking facilities are available and alternative
padding materials could be adopted. The estimated cost
is about three times that of the traditional yoke. The
same principle of connecting two collars to form a double
yoke could be applied to the other single collar harness
designs described. It is understood that this type of
harness has also been used experimentally in Madagascar.

Double Harnesses - Attachment of Implements

The attachment of short and long beam agricultural implements is essentially the same as for traditional double yokes. However it should be noted that on some of the improved harnesses the hitching point is lower than on the traditional yoke.

For short beam implements the connection is by means of a rope or chain from the implement to the hitching point at the centre of the harness. The hitching arrangement can be designed to compensate for differences in draught output between two animals while maintaining an even pull on the draught chain. In the arrangement shown opposite the length of the links between the yoke and the draught chain can be altered to accommodate animals of different strengths.

For long beam implements the beam is attached directly to the hitching point at the centre of the yoke. The yoke should be free to pivot about the axis of the beam to allow relative vertical movement of the animals when passing over uneven ground. Allowing the yoke to pivot about the vertical axis makes manoeuvring and turning easier but according to some experts, can cause problems if the two animals have significantly different power outputs or natural working speeds. Illustrated opposite is a technique used in Mali which allows the yoke to pivot about the axis of the beam but prevents movement about the vertical axis.

The attachment of a two-wheeled cart or other wheeled implement to a double yoke uses the same arrangement as for other long beam implements. That part of the cart load supported by the animals is applied to the neck. Reversing and braking capability can be improved by the addition of a breech strap, as described in the section on Single Harnesses, to prevent the yoke sliding forward along the neck.

Yoke viewed from above

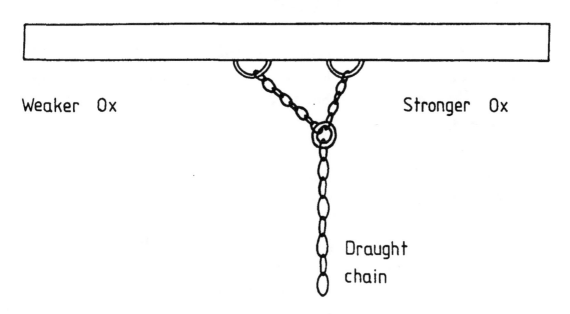

Weaker Ox   Stronger Ox

Draught
chain

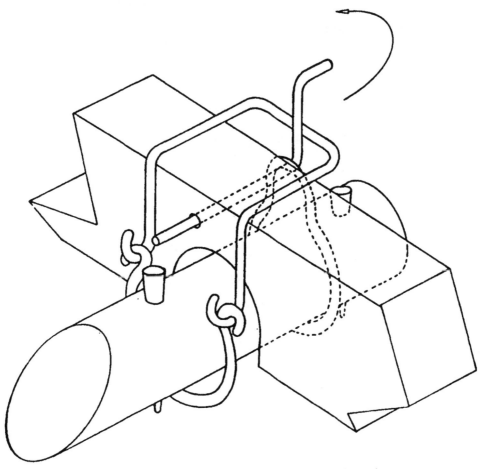

Centre section of yoke

Independent hitching of pairs of bovines

As noted earlier independent hitching of bovines is not common at present. However the technique is commonly used with equines and the different methods are described in Section 3.3 of this report.

The reader should therefore refer to Section 3.3 for details of the methods used.

The independent hitching of a pair of bovines to an implement is exactly the same as for a pair of equines, i.e. through two swingle-trees and an evener bar.

No evidence has been found of hitching a pair of bovines to a wheeled implement using a dissel boom as described in Section 3.3. However in China the arrangement is used where one animal both draws the implement and supports the shafts, while the second animal simply draws the implement.

PART 3: EQUINE HARNESSES

The main draught strength of equines lies in the shoulders and breasts. This is in contrast to bovines which are weak-breasted. Further, the physiology of equines is such that it is acceptable to have load bearing elements of the harness passing across the breast in front of the animal's forelegs. As a result of the physiological differences between equines and bovines:

(i) neck yokes are not suitable for equines

(ii) two types of equine harness have evolved:

   - the breastband harness
   - the full collar harness

The breastband harness is the simpler, and cheaper type, and is satisfactory for light duty work. The full collar harness is preferred for heavy duty work. Both arrangements consist of a basic draught harness together with additional elements that are used for pulling different types of implement. Where two or more equines are needed to provide sufficient draught effort, it is normal to use independent hitching rather than to link the animals together by a rigid harness.

Information on equine harnesses is presented here in three parts:-

-    basic breastband harnesses
-    basic collar harnesses
-    methods of hitching implements to one or more equines

3.1 BASIC BREASTBAND HARNESSES

A typical, basic breast-band harness is illustrated opposite.   Its major elements are:-

- a broad strap passing around the animal's chest through which the draught force is applied, and at the ends of which the attachment points for the traces are fitted - these attachment points might be rings, hooks or holes:
- an adjustable shoulder strap which positions the breast strap correctly on the animal.

The crucial requirements of the breast-band harness are that:

- the main breast-strap must be fairly wide so that the draught force is spread over a reasonable area of the chest, and must be a comfortable fit on the animal:
- the broad strap should lie across the animal's chest, and not ride up against the throat where it would cause the animal to choke:
- the broad strap should not interfere with the free movement of the fore-leg joints:

Certain adaptions or improvements of the basic breast-band arrangement are illustrated overleaf:

1. The shoulder strap may be forked to give better location of the harness on the animal.

2. The shoulder strap can be padded at the top to distribute the load imposed on the animal.

3. The strength of the breast strap may be increased

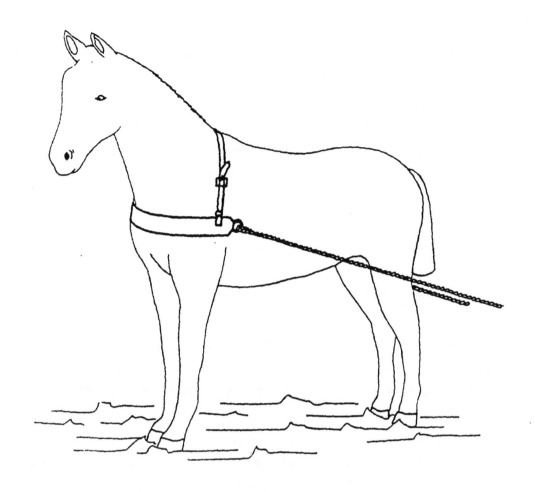

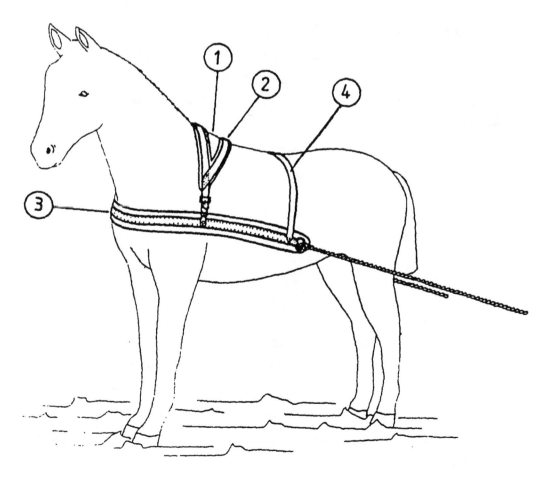

by making it in two parts, a broad inner strap which spreads the load over the animal's chest and a narrow outer strap which transmits the draught force to the traces. The outer strap is slightly shorter than the inner one which therefore protects the animal's body from rubbing by the trace attachment.

4. The breast band can be extended, and further located by a back-strap, to give improved guidance of the traces.

Breast-band harnesses in developed countries are normally made of leather. The elements of the harness are either directly stitched together or joined by rings stitched to the straps - stitched construction is considered to be preferable to rivetting. Leather is expensive and/or difficult to obtain in some countries and its processing requires considerable skill. A further problem with leather is that it requires regular maintenance - washing with soap and water and oiling - to remain soft and supple, particularly in hot, humid climates, and if it is subjected to perspiration. It is thus worthwhile considering alternative materials for the harness.

Detailed on the following pages are attempts to find alternatives to leather for breast-band harnesses.

1. Rope

In Niger, a breast-band harness for donkeys has been used with the straps made from ropes spliced together. This is preferable to tying the ropes since the resultant knots would tend to rub against the animal. The ropes were padded in the critical areas where they press against the animal. This form of construction was adopted since the local leather was too weak to be used in a load-carrying capacity. However it proved to be an acceptable padding material.

2. Rubber

Rubber breast-band harnesses for <u>donkeys</u>, made from old car tyres, have been identified in Botswana and Tanzania. The Botswana design is illustrated opposite.[1] Note that it has a shoulder strap and a back strap. The straps are cut from the casing of old car tyres using a sharp knife, and any excess tread rubber removed. These straps rely on the ply reinforcement of the tyres for their strength. The straps are "stitched" together with thin wire. Care must be taken to ensure that the stitching does not rub against the animal and cause sores. The designers recommend that the breast strap be wrapped with burlap sacks or other cloth. These absorb perspiration and are a better material to have next to the skin than rubber. The traces, straps cuts from old car tyres, are attached to buckles on the breast band.

The Tanzanian rubber harness illustrated opposite is again designed for donkeys and uses straps cut from the tread of an old car tyre. The harness has only one locating strap but this is set well back and is in effect a back rather than should strap. The two straps are joined together by piercing holes in the straps with a chisel and then tying them together. Additional holes are pierced at the ends of the breast strap for attachment of the traces. The harness has two additions to the basic breast-band arrangement:

- a girth, or belly, strap tied to the harness

- a rope tied from the front of the breast-band around
  the neck. This prevents the breast-band from slipping
  downwards.

It is recommended that the breast, back and shoulder straps be padded by binding them with sacking.

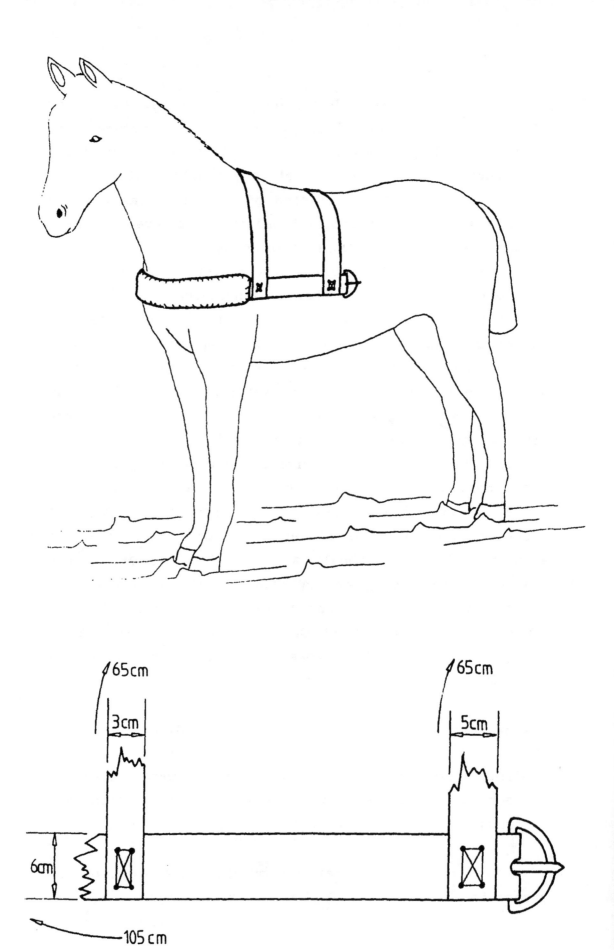

65cm

3cm

65cm

5cm

6cm

105 cm

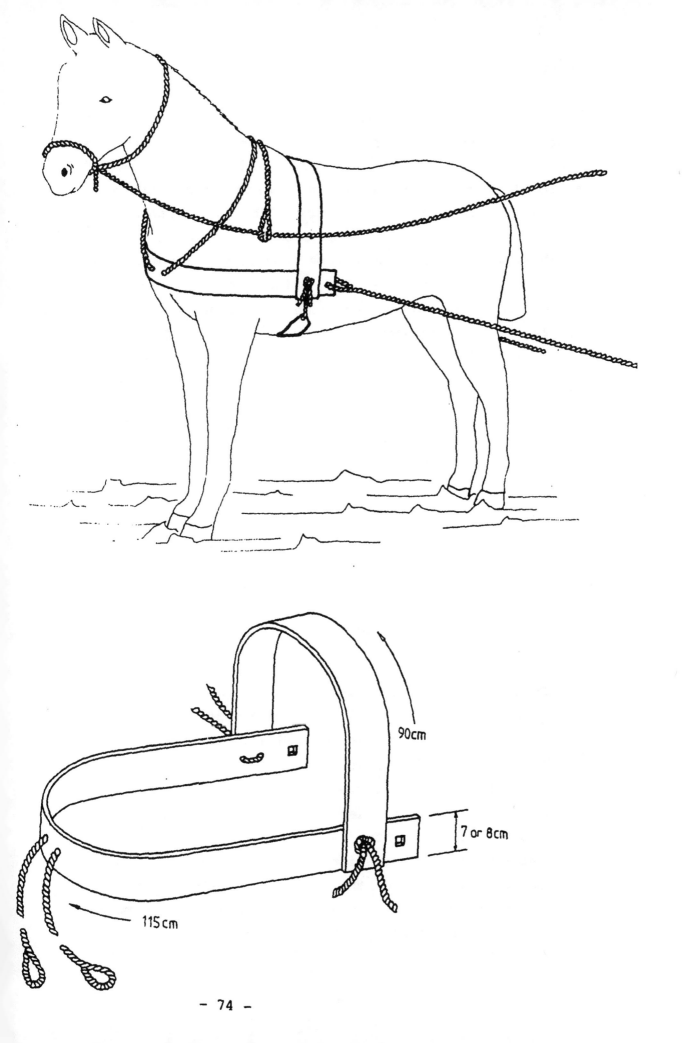

90cm

7 or 8cm

115cm

3. Webbing

In Botswana, breast-band harnesses have been made from woven webbing/canvas straps. The overall arrangement is the same as for the Botswana rubber harness and is of stitched construction with the joints reinforced with leather. Initial testing showed that the harness was strong and effective but no data is available on its durability. The webbing straps require washing with soap and water from time to time, since they will tend to harden if they absorb too much perspiration. This problem could be reduced by padding the breast strap.

[1] Where available, dimensions are given for the harnesses described here. These dimensions are for guidance only and should be used with caution. There are considerable differences in size and strength of different breeds of donkey and horse which will affect the length and cross-section of straps required. Some care has been taken in preparing the illustrations to show accurately the positions of the straps on the animals. These can therefore be used as a guide in measuring animals to define the dimensions of the harness.

3.2 COLLAR HARNESSES

The basic, traditional European collar harness consists of two major elements, the collar and the hames:

> the collar is a full collar fitting around the base of the neck and passing across the breast and in front of the shoulder joints. To avoid causing injury to the animal the collar must be a correct fit around the animal neither too loose, which causes rubbing and sores and may interfere with free movement of the fore-legs, nor too tight, causing pinching. The collar is made of leather, and is padded.

> the hames fit into a slot formed in the periphery of the collar. They are two curved metal pieces joined by a chain at the bottom and an adjustable strap at the top to ensure that the hames fit correctly into the collar. The trace attachment points are fitted to the hames slightly above the joint of the animal's shoulder blades. In addition to the collar and hames, a back and belly strap is sometimes fitted to assist in locating the harness and guide the traces.

The hames provide the necessary rigidity for the harness, and transmit the draught force to the traces. The collar ensures that the draught load is applied without discomfort to the animal and is distributed across the breast and the shoulders. The production of a good quality collar requires considerable skill, for it is a complex example of the harness makers art, and is expensive. The best quality collars are purpose-made to fit a particular animal. It is also important that the hames are a good fit in the collar. Finally, the collar requires regular maintenance - washing and oiling - to keep it in good, supple condition.

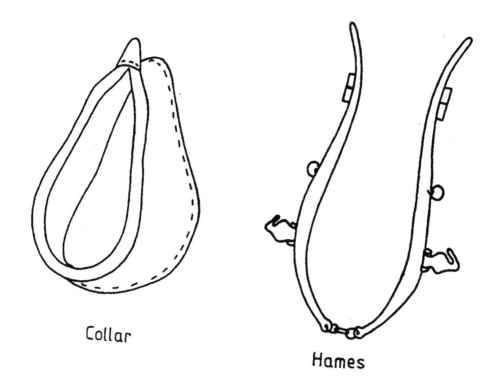

Collar

Hames

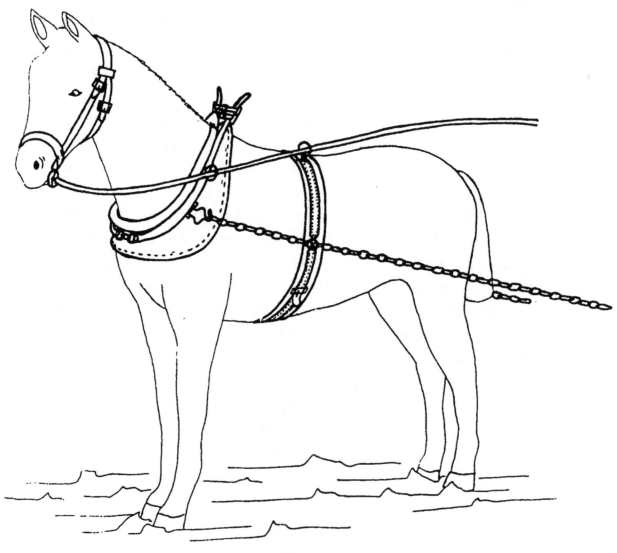

The complexity of manufacture, the cost and the maintenance requirements of the traditional horse collar mean that it is inappropriate to many developing country situations.

Two approaches have been identified to provide simple, cheaper collar harnesses more suited to developing country conditions.

Wooden-framed collar harness

Several collar designs have been developed in Africa, all consisting essentially of a two-part wooden frame (the hames), padded to provide a comfortable fit and to distribute the load over the animal's body. Most of this work has been concerned with providing suitable harnesses for donkeys, though the same principles could be applied to horse collars.

The critical requirements of these wooden framed collars, if they are to be effective, are:-

- that they should be a good fit on the animal
- that there should be sufficient quantity of padding to distribute the load and prevent the frame pressing against the animal
- the frame should be rigid, and not flex when a draught load is applied to the traces, since this will cause the animal discomfort and may impede free movement of the shoulders

Three versions of the wooden framed collar harness are illustrated.

1. The two parts of the wooden frame are joined by a bent metal strip attached to the inside of each frame. The frame is joined by a strap at the top and is adjustable at the bottom to provide a good fit on each animal. Padding is fitted to the inside of each frame. Attachment rings for the traces are fitted at approximately the mid-point of each frame.

2. The second example is similar to the first, but in addition to the bent metal strip has a wooden bracing piece joining the two frames at the top, and

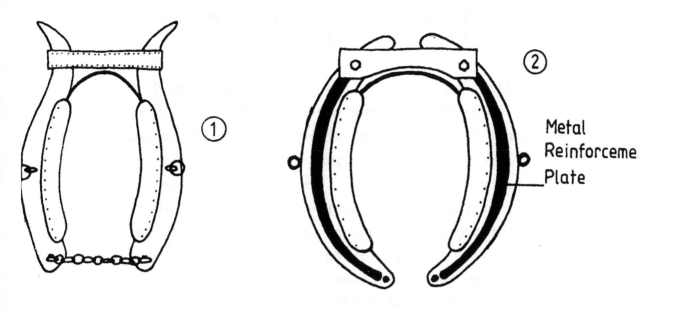

①   ②

Metal
Reinforceme
Plate

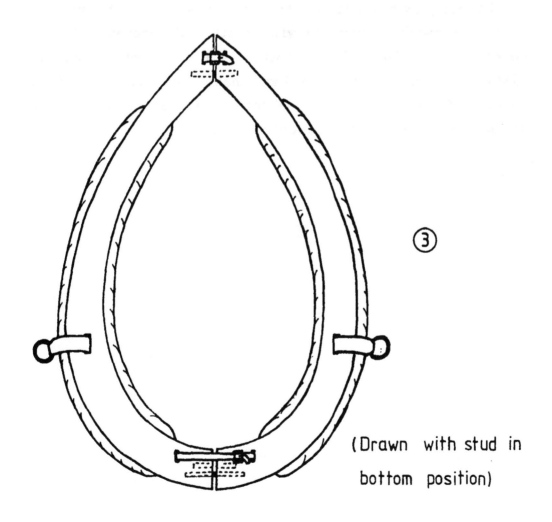

③

(Drawn with stud in

bottom position)

metal strengthening pieces on each frame. These
measures increase the rigidity of the harness.

3. The first two examples have the padding fitted to
    the inside of the frame on each side. The third
    example has the padding at the back of the frame,
    and covering a greater area to produce what is close
    to being a full collar. The rigidity of the two
    parts of the frame is achieved by fitting a metal
    stud at top and bottom into matching holes in each
    part of the frame. Straps are used to tighten
    the frames up against the studs. There is a
    choice of holes, of different depths, for the bottom
    stud which provides adjustment of the width of
    the harness.

Padding materials quoted for the above harnesses
include horse-hair covered with cloth and vegetable fibres
covered with plasticized cloth. Plasticised cloth is
likely to be harder wearing than material made from
natural fibre but unless selected carefully, may cause
damage to the skin where rubbing occurs.

Car-tyre collar harness

In Botswana experiments have been carried out in making collar harnesses from old car tyres. The idea is based on the fact that, if the sidewall is cut from the tyre, and positioned so that the inside of the tyre faces upwards, it adopts a shape similar to that of a collar. Two designs were tried out, and the more successful is illustrated opposite and described below.

The collar is made in two halves, each of two layers of tyre. The two layers and the two halves are joined by stitching with thin wire or by rivetting. An adjustable strap at the top of the collar allows it to be fitted to a particular animal. The underside of the collar is fitted with cushion pads. Dried grass stalks and strips of burlap were tried as padding materials and the former was found to have better perspiration-absorbtion properties. The harness is completed by an adjustable girth strap joined to the collar by side straps all cut from car tyres. The traces are attached to the ends of these side straps.

In tests on a labour-intensive road construction project in Botswana one major problem was identified with this harness. Because the collar is not rigid it distorts when the draught load is applied to the traces. This caused the wire joints and rivets to break and brings the sharp metal parts and the edge of the rubber collar into contact with the animal, leading to abrasion sores. It was also found that a substantial amount of padding was required to prevent the metal parts contacting the animal. However, it seems likely that, with the addition of rigid hames, which could be made from wood, the concept provides the basis for developing an effective collar harness.

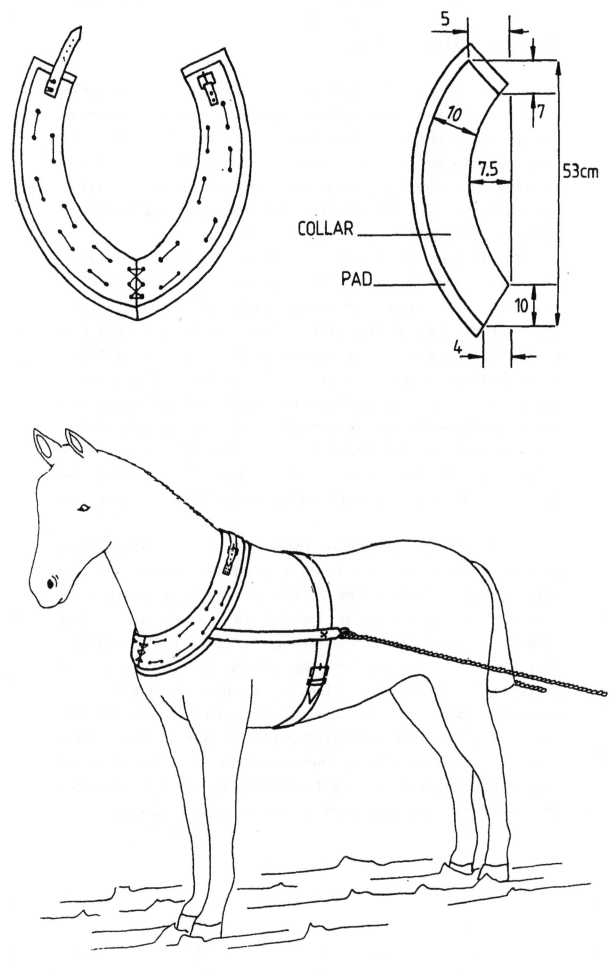

COLLAR

PAD

5

10

7.5

7

53cm

10

4

3.3 EQUINE HARNESSES - ATTACHMENT OF IMPLEMENTS

The arrangements for attaching implements are essentially the same for both breast-band and collar harnesses.

Single Animal

The principles of attaching a single equine harness to implements are the same as for bovines.

For agricultural implements the traces are attached to the ends of a swingle-tree at the rear of the animal and the implement is connected to its centre. Some clearance must be allowed between the rear of the animal and the swingle-tree which must be sufficiently wide to prevent the traces rubbing against the animal (see Section 2.3). Some of the breast-band and collar harnesses described include a girth or back strap which assists in guiding the traces but this is not essential. The traces can be made of rope, leather or chain. As discussed for bovines 'closed-link' chain is preferred.

Wheeled implements, including two-wheeled carts, are attached to a single equine harness by means of two shafts running forward from the implement one on each side of the animal. The standard practice is for the shafts to be attached by means of a saddle fitted on the animal's back. This is the preferred method for equines since they will accept a load being applied to the back.

In its simplest form the saddle can be a broad belt, held in place by a belly strap, with a loop on either side into which the shafts fit. An alternative arrangement is to use a wooden saddle, either padded or sitting on a saddle cloth, with a strap passing over it with loops at each end. Again the saddle is held in place by a belly strap. The traces are attached from the harness either to the shafts or directly to the cart. The length of the traces and the shafts should be such as to prevent the

latter slipping backwards out of the loops, while
providing clearance between the rear of the animal and the
implement.   To facilitate braking and reversing, and to
prevent the cart running forward into the animal, a breech
strap should be fitted.

Note:the straps can be made of leather, webbing, or strips
cut from old car tyres.   The breeching strap can be
made from rope, but care should be taken to prevent
it causing sores on the animal, particularly where
knots are tied.

Note:see also Section 2.3 for discussion of attachment of
wheeled implements to a single animal harness.

Two animals

As noted earlier, where two equines are used to draw an implement, they are fitted with single harnesses and independently hitched to the implement.

The arrangement for hitching two equines to an agricultural implement is as follows:

Each single harness is connected by a pair of traces to a swingle-tree at the rear of the animal. The two swingle-trees are joined by an evener, to the centre of which the implement is attached.

This arrangement allows two animals of different strengths to work side by side drawing the implement. The lever arms on the evener (i.e. the distance between the implement and swingle-tree attachment points) can be adjusted so that the two animals give equal pull on the implement.

The normal arrangement for hitching two equines to wheeled implements, including two wheeled carts, is by means of a "dissel boom" in conjunction with collar harnesses. The dissel boom is a rigid shaft which passes forward from the implement between the two animals. The animals are hitched by means of traces and swingle-trees to an evener bar which is pivoted at the rear of the dissel boom. There are three possible means of connecting the front of the dissel boom to the harnesses as described overleaf.

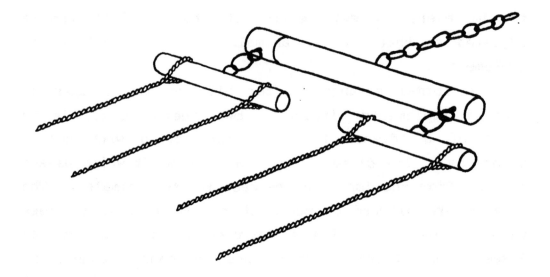

1. By means of two chains, connected to the base of
    each collar, from the front of the dissel boom.

2. By means of a crossbar which can twist and rotate
    relative to the dissel boom and which is connected
    at each end by chains to the base of each collar.

3. By means of a crossbar connected as above and two
    swingle-trees are attached in turn by pairs of
    straps to the girth strap of each harness.

The first method is the simplest, but the second is
preferred since it keeps the collar better located. The
third method is probably the most efficient, though the
most complicated, since it means that any vertical load on
the dissel boom is applied to the animals' backs rather
than to their shoulders.

In designing the above arrangements care must be taken
in setting the lengths of the traces and the dissel boom
to ensure free movement of the animals' fore-legs.

There is a second method of hitching two animals to a
wheeled implement. This is to use exactly the same
arrangement as for a single animal (i.e. with a pair of
shafts and a saddle) and to hitch a second animal, positioned in front of the first, directly to the implement by
means of a pair of traces. In this arrangement the first
animal both pulls the implement and supports any vertical
load applied through the shafts, while the second animal
simply pulls the implement.

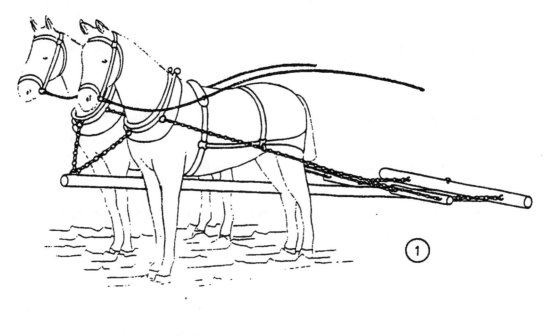

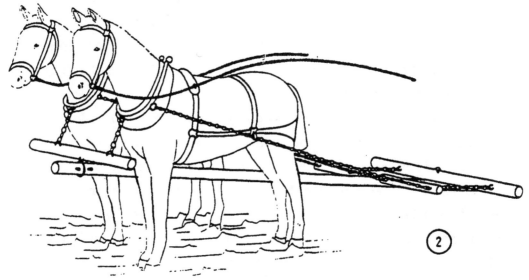

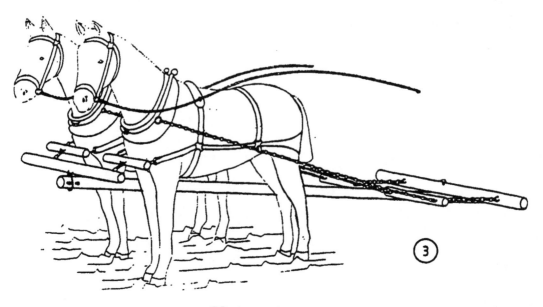

PART 4: INFORMATION SOURCES

This is a limited selection of easily available
publications containing useful information on the
harnessing of draught animals.

DEVNANI R.S. "Design considerations for harnesses and
yokes for draught animals". Central Institute of
Agricultural Engineering, India. December 1981.

available from: Central Institute of Agricultural
                Engineering,
                Nabi Bagh,
                Berasia Road,
                Bhopal, 462010,
                INDIA

*   F.A.O. "Farming with animal power" Better Farming
    Series No. 14. FAO. Rome, 1977.

*   F.A.O. "The employment of draught animals in
    agriculture". FAO Rome 1972.

GOE M.R. & R. McDOWELL "Animal traction: guidelines for
utilisation". Cornell University, New York, 1980.

available from: Department of Animal Science,
                New York State College of Agriculture
                & Life Sciences,
                Cornell University,
                Ithaca,
                New York, 14853,
                U.S.A.

* HOPFEN H.J. "Farm implements for arid and tropical regions". FAO Agricultural Development Paper No. 91. FAO Rome 1969.

* RAMASWAMY N.S. "Report on draught animal power as a source of renewable energy" prepared for U.N. Conference on New and Renewable Sources of Energy. FAO Rome 1981.

SWAMY RAO A.A. "Report on the preliminary investigations, design and development, testing and economic analysis of the new single and double bullock harnesses at the Development Centre from May 1962 to April 1964". Allahabad Agricultural Institute, India 1964.

The original report is very difficult to obtain but a synopsis is available from:

> Intermediate Technology Transport Ltd.,
> Home Farm,
> Ardington,
> Oxon. OX12 8PN.
> U.K.

WATSON P.R. "Animal Traction" Peace Corps, Washington 1981.

available from: Peace Corps,
> Information Collection & Exchange,
> Office of Programming and Training
> Co-ordination,
> 806, Connecticut Avenue, N.W.
> Washington, D.C. 20525,
> U.S.A.

U.S.A.I.D. "The buffalo as a draught animal in Thailand". U.S.A.I.D. 1979

available from: The Office of Agriculture,
                U.S. Agency for International Development,
                320, 21st Street, N.W.,
                Washington, D.C. 20523,
                U.S.A.

*   F.A.O. Publications are available from:

                Publications Division,
                Food & Agriculture Organisation of the
                United Nations,
                Via delle Terme di Caracalla,
                00100 Rome,
                Italy.

www.ingramcontent.com/pod-product-compliance
Lightning Source LLC
Jackson TN
JSHW060311140125
77033JS00022B/645